Tips for Engineering Students:

The best student engineering tips to get you through college or university

Tips for Engineering Students:

The best student engineering tips to get you through college or university

Dylan Thomas

Table of contents

Table of contents	5
Introduction	9
1. Get a reliable alarm clock	11
2. You'll have to rely on yourself and use the initiative	12
3. Plan your note taking (paper vs digital)	13
4. Buy used and old edition textbooks (if needed)	15
5. Join at least one club or organisation outside of engineering	16
6. Don't be afraid to speak out if you're struggling	17
7. Stay on top of lectures and work	18
8. Complete work when you get it	19
9. Eat enough good food	20
10. Learn to use your calculator properly	21
11. Pomodoro method of studying is a game changer	22
12. Don't be afraid to ask questions	23
13. Work on your teamwork	24
14. Identify your mentor	25

15. Find friends on your course to help study with — 26

16. Keep involved in the real engineering world — 27

17. Seek engineering challenges to get involved with — 28

18. QUIZ BREAK — 29

19. Don't pause your professional development — 30

20. Seek a summer internship or year in industry — 31

21. Develop your goals — 32

22. Continuous flash cards — 33

23. YouTube is a goldmine — 34

24. Take a business class — 35

25. Attend those tutor sessions — 36

26. Learn what interests you — 37

27. Work on what you struggle with — 38

28. Don't always study alone — 39

29. Make study "cheat" sheets for your exams — 40

30. Everyone in your class feels the same as you — 41

31. Try to attend all lectures, but it's not the end of the world — 42

32. Don't be competitive or toxic — 43

33. It's normal to fall behind a bit — 44

34. Get on LinkedIn and use it — 45

35. There's a lot to be learned online — 47

36. It's sometimes possible to find PDFs of textbooks online — 48

37. Create a portfolio of your work — 49

38. Ask for help when you need it — 50

39. WORD SEARCH BREAK — 51

40. Learn and work on other soft skills — 52

41. There's more to life than being number 1 — 53

42. Become S.M.A.R.T — 54

43. Learn to code! — 55

44. Get hands on — 56

45. Attend those networking events and job fairs — 57

46. Apply for jobs early — 58

47. Practice good time management — 59

48. You won't use everything you learn — 60

49. Be proactive — 61

50. Improve your communication skills — 62

51. Work on your strengths — 63

52. Look out into the world of engineering — 64

53. Nobody is perfect, don't beat yourself up   65

54. Don't be THAT engineering guy   66

55. Try and create a study timetable during exam season   67

End of guide   69

Introduction

Hello and thank you for choosing this guide to help guide you through your time studying engineering!

The aim of this guide is not to be a replacement engineering textbook with technical details, but to generally give some advice about helping you survive your time while studying.

The best way to use this guide is to check in every so often and just take on board a few of the tips at a time. I'll be impressed if you managed to remember all the tips at once!

Whether you're thinking of studying engineering, or maybe you're in your final year, whoever you are I'm sure you'll find this useful.

1. Get a reliable alarm clock

Making sure you have a clock that gets you up in the morning for lectures and class is one of the most important things you can get! The amount of times that I missed class because my phone ran out of battery or I forgot to set my alarm would surprise you. I know phones these days are pretty reliable, but sometimes if you're out the night before and you know you need to be up, having that alarm already ready back at home is always a lifesaver.

2. You'll have to rely on yourself and use the initiative

College and University is a lot different to how you've been taught normally throughout your life up to this point. You'll be taught in bigger groups in lectures and will be taught the general requirements of what you need to know, but there will be a lot of work that won't be taught and will require a lot of self determination. For example, my lecturer told me that for every hour I had a lecture, I should be doing 7 hours of my own work! That's a lot of time spent by yourself working to get the results you need. Now, obviously the lecturer is always going to exaggerate, I'm not sure I did 7 hours per lecture. The point however is that a lot of work is required other than just attending lectures!

3. Plan your note taking (paper vs digital)

Before you begin is the time to decide how you want to take notes, electronically or physically with paper. It'll be very difficult to switch back to a different method once you begin! I would say there are perks and downsides to each method which I've shown below:

Method	Pros	Cons
Electronic	• Much easier to search back through old documents to find what you're looking for. • Can take physical pictures of board and add them directly to notes. • Can link directly with lecture slides if accessible online. • Can make the notes appear more visually appealing when compared to paper.	• Can be expensive initial set up cost to get good equipment (iPad, laptop). • Can be slower to take notes and draw diagrams when following along with lecturer. • Can take a bit of time to set up get ready at start of lectures. • Things need charging!
Paper	• Faster set up and ability to take notes. • All you need is a pen and paper. • Much easier to just draw diagrams to follow along with lecturers.	• Requires good planning and filing to make sure you won't lose your notes. • Creates a lot of paperwork that needs organising.

I personally used paper during my time at university, with an expanding folder to store my notes during the semester with a different pocket for each module, which I then transferred to binders and folders at the end of the terms.

4. Buy used and old edition textbooks (if needed)

Textbooks can be really useful, but sometimes it can be daunting when the lecturer shows 10 textbooks that you're required to buy straight away, and you need the latest version and each one costs over £200/$200! Now sometimes some of them are actually required for the course, but it might not be required to get the latest edition. Older textbooks on eBay go for way, way cheaper. If it gets a few weeks into the course and you definitely need the latest version, see if you can share with some other people to spread the cost. Some of the textbooks aren't even needed sometimes.

Mechanics of Fluids
Bernard S. Massey
★★★★☆ 12
Hardcover
£175.28
Get it 11 – 15 Aug
Remove from view

VS

Mechanics of Fluids
› B. S. Massey
★★★★☆ 13
Paperback
15 offers from **£0.70**

5. Join at least one club or organisation outside of engineering

University and College has a lot more to offer than just the course you're doing. There are societies, clubs and organisations that cover everything you can imagine. Make sure you get involved with some of these clubs outside of your normal study. These may be related to engineering if that's what you're into, but don't be afraid to join some random clubs or sports you like the look of. Don't be afraid of what it will look like on your CV when you apply to jobs, it's really important to make sure you find something you're interested in as it will help your mental health a lot more and let you grow as a person.

6. Don't be afraid to speak out if you're struggling

Engineering is a very difficult course, there's no denying that. Because of this, make sure you look after yourself and know that there are always people you can talk to or people that can help if you ever feel like you're struggling, even if you think it feels silly, I can guarantee that it's not. A lot of times, your own universities will have support numbers and people you can speak to if you feel like you're struggling, or even reach out to a friend to chat to, don't struggle alone!

7. Stay on top of lectures and work

This one is a lot easier said than done, especially when work begins to pile up. However, it's a really important one and one that you should strive to do at all times. You will receive a lot of content in lectures, and taking some time after a lecture to go over your notes and check that it makes sense and writing a summary of what's important to remember will come in very handy for exam time. Even if it's just 15 minutes per lecture, future you will absolutely be grateful. This can also be an effective time to make flash cards if that's how you study, it's already fresh in your mind.

8. Complete work when you get it

Making sure that you complete coursework or homework when you get it is also a lot easier said than done, however, it will make your life so much easier. Work will very quickly pile up if you don't complete if when you get it so doing it when you get it will help a lot. Even if you don't manage to complete it, making a start to the work as soon as possible and planning out what you want to write or how you're going to complete the work will help you tons when you actually come around to finish the coursework as it's all already planned. It helps with the stress as well knowing that you've already begun work or even completed work, you can relax a bit more.

9. Eat enough good food

So, for a lot of people this will be the first time you've ever lived by yourself, and that means for a lot of people cooking and buying food for yourself for the first time (not everyone though I know). It's so important to have good food and enough of it. Just because you might be on a low budget doesn't mean you have to eat rubbish food! Cooking is not as hard as some people make it sound! Find a simple cookbook or online website with very easy recipes, and follow it precisely. Within weeks you won't even need a cookbook and you can make your own meals. It's hard to think straight and work hard when you're running on no food or rubbish food.

10. Learn to use your calculator properly

It is amazing how many people don't even know what their calculators are capable of, don't be one of them! Calculators that are allowed in exams have so many functions that people don't realise, that help out so much during exams and when doing work. Solving equations, memorising numbers, storing constants and converting units just to name a few. It's so important to learn how your calculator works and make full use of it. You can do this by reading the manual, or just going online and seeing what your specific calculator can do. Make sure you get the best one that's approved for use in exams, every little helps!

11. Pomodoro method of studying is a game changer

Now I think we can all agree, studying is hard, especially with the wealth of distractions that we have at arms reach on our phones and laptops. I know that I'm guilty of it which is why the pomodoro method of studying helped me so much during my studying time. It works by breaking down your work into chunks of 25 minutes, separated by 5 minute breaks. After 4 pomodoros, take a longer break of 30 minutes. This lets you focus for your time in the 25 minutes then you can go back on Instagram or YouTube for those 5 minutes, then it restarts. You can even get browser extensions to force it that block websites for the 25 minutes, and open them for 5 minutes. It's definitely worth trying if you struggle to focus and find yourself drifting off and using your phone or other websites.

12. Don't be afraid to ask questions

Asking questions is very important but also very hard, during lectures I always struggled to ask questions because I was afraid people would judge my questions. Once I realised that everyone else wanted to ask the exact same question but was afraid as well, I just asked every chance I got. Believe me, everyone will be grateful. It's not just lectures where you can ask questions as well, send emails to lecturers, go to study sessions or ask friends for help. A lot of times people will always be willing to help you out. Make use of all the sessions you can.

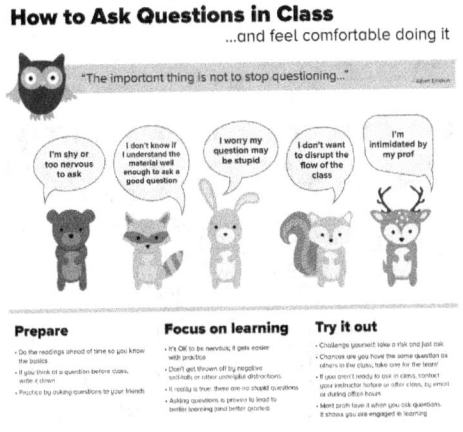

13. Work on your teamwork

Teamwork in college and university is a struggle, I've been there and I know it can sometimes be like pulling teeth. However, it's such an important skill to have to be able to work effectively with other people who you've never spoken with before. There will be people who don't pull their weight and who you don't get along with and you'll be forced to work with them. Just get your head down and work with them as best as you can to get to the end goal. If certain people aren't pulling their weight, give them a few chances before going to the lecturer, but sometimes it's necessary. In our courses we could rate each team members performance anonymously which worked pretty well at calling out the slackers.

14. Identify your mentor

Having a mentor to help you in your studies is really useful, and I wish I made better use of mine when I had the chance. Personally, I had one assigned to me, but if you don't have one given to you, don't be afraid to ask some lecturers if they would help mentor you. They don't even need to be your lecturers, they can be members of the university you just come across. They will give you great advice of working, helping with questions you have, looking over CVs, interview practice and just generally helping out if you need a rant (maybe about those team members who slack).

15. Find friends on your course to help study with

Finding friends can be hard, but remember that everyone else is in the same position as you on your course, looking for friends! It's a great time to meet new people and find friends. Friends and people on your course will always come in handy when you're stuck on work or when you want to ask some questions, it's important as well to help them out as well when they need help. We set up group chats on my course where we would have a load of people on the course asking and answering questions, which was so useful. Don't be a freeloader though and not help!

16. Keep involved in the real engineering world

Keeping up to date with what happens in the engineering world isn't a requirement, but it's good practice and it lets you stay up to date with everything going on in the real world and lets you see how engineering is put into practice. Staying up to date with all the latest things will only benefit you later in life and is a good habit to form, finding those Twitter accounts or websites which tell you the good stuff will always come in handy. It also just helps to show that you're a good all rounder and you're not just in your student bubble with no understanding of the outside world!

Discipline	Source of information
Aerospace	Aviation Week & Space Technology
	Flight International
	space.com
Chemical	Chemical & Engineering News
	Oil and Gas Journal
Electrical	EDN
	IEEE Spectrum
	New Electronics
Mechanical	Energy and Environment
	Journal of Fluid Mechanics
	Engineering and Technology Magazine

17. Seek engineering challenges to get involved with

There will no doubt be some engineering challenges at your place of study that allows you to get involved with some engineering challenges. These may be something like Formula Student, or maybe solving some challenges for places in need of improvement or even a rocket building team. Seek out these challenges and get involved! It will help you develop those engineering skills and actually put them into good use. Applying those engineering lessons you learn will help embed them in yourself. You can even just build something yourself to challenge yourself like a website or a drone. Embedding those skills will only improve yourself in the long run.

18. QUIZ BREAK

1. What is the ratio of stress to strain?
2. What is the unit of work?
3. What is the universal gas constant in kJ/mol•K?
4. What is the first law of thermodynamics?
5. What is the definition of work done?
6. What are the 4 main cycles of an internal combustion engine?
7. What is laminar flow?
8. What are transformers used for in electricity?
9. When a fourth resistor is connected in series with three resistors, the total resistance…
10. What material would you use in the hot end of a jet engine?

1. UTS 2. NM 3. 8.314 4. ENERGY CAN BE TRANSFORMED FROM ONE FORM TO ANOTHER, BUT CAN BE NEITHER CREATED NOR DESTROYED. 5. W = FS 6. SUCK SQUEEZE BANG BLOW 7. FLUID PARTICLES FOLLOWING SMOOTH PATHS IN LAYERS, WITH EACH LAYER MOVING SMOOTHLY PAST THE ADJACENT LAYERS WITH LITTLE OR NO MIXING. 8. TRANSFERS ELECTRICAL ENERGY FROM ONE ELECTRICAL CIRCUIT TO ANOTHER CIRCUIT, OR MULTIPLE CIRCUITS 9. INCREASES 10. NICKEL-BASED SUPERALLOY OR CERAMIC BLADES

19. Don't pause your professional development

Professional Development is an important skill to continually record and track your development through your career. You may think that because you're so early in your career it might not be relevant to you, but it is the best time to begin and form good habits.

One of the best ways to track your professional development is by using SMART. You may have come across this before but it stands for SPECIFIC, MEASURABLE, ACTION, RELEVANT, TIME. I talk about this a bit later in the book. The idea is to continually record your achievements and make them relevant to your goals and plans for your career, and store them with CVs, courses and other important documents. This makes your career development a lot easier to track and helps you a lot when it comes to interviews, jobs and accreditation.

20. Seek a summer internship or year in industry

The opportunity to undertake a summer internship or year in industry is not one to be ignored. The ability to work in industry for 3-12 months while gaining a wealth of knowledge and experience is definitely hugely beneficial to yourself but also your career. I was fortunate enough to do a year in industry and gained a huge amount of knowledge which immediately jumped me up the ladder when it came to applying to jobs and graduate schemes. It also lets you apply the knowledge you learn in your studies to the real world, and see how these companies function in the real world.

(For people in the UK, I would highly recommend using etrust.org.uk)

21. Develop your goals

Having developed goals is always easier said than done, I used to struggled when people asked "where do you want to be in 5 years" and the truth was I didn't know, and still know I'm not 100% sure. But it's important to set general goals of where you'd like to end up so you can tailor your development and skills to meet these goals. For example, do you want to become highly specialised and technical in your field, or maybe you would like to be a generalist with a good understanding of a lot of areas. Would you like to become a manager of people or do you see yourself as an engineer forever, or at least the next 5 years? Do you have specific interest? Do you want to travel, move abroad for some time? Answering questions like these helps to develop your goals and helps you develop yourself.

22. Continuous flash cards

Depending on how you learn, making continuous flash cards can be a life saver. If you learn some things using flashcards, it's a great idea to continually make them after lectures and in your free time and find some random time to study them. When it rolls around to exam time, most of your flashcards are ready made to be learned.

23. YouTube is a goldmine

YouTube helped me so much with my degree, and it will definitely help with yours. Sometimes the lecturer or professor will explain a certain topic that may not make sense to you, and no matter how many times you read it and try and understand it, you'll still struggle. What you need in times like these are to hear the problem explained from a different perspective, which is where YouTube comes in! There are a lot of resources out there about almost every single topic on Engineering, so if you're ever struggling, make sure to check on there and hear it explained from a different perspective.

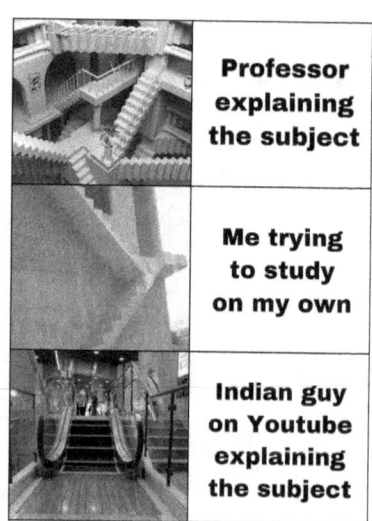

24. Take a business class

Having a broad understanding of the business world and not just engineering is really useful. Understanding how a business works and makes money is actually quite important and you'd be surprised how many people out there don't know about it. You may think you don't need to know how it works, but chances are the company you go and work for exists to make money, and understanding how they do it and what that really means can be very beneficial. I'm not suggesting you should take this class to become some start up and start your own business (although that is definitely a possibility), but just to broaden your understanding of the world and learn more about how things work.

25. Attend those tutor sessions

It's likely that you'll have tutor sessions available to you from your professor to go through certain questions, topics and to just make time available to help you with your studies. Don't skip them! They can be more useful than some lectures as you're applying the knowledge directly into questions which could well very likely come up in your exams. Make sure to ask lots of questions and get involved with what the tutors are saying. It may even be useful to have some questions prepped and ready to go before the sessions to be answered which will help your understanding greatly.

26. Learn what interests you

Learning and focusing on your interests really helps your through your degree as it's really useful to come back to topics that interest you and maybe you understand more than others. There will always be topics which are very difficult and sometimes make no sense, but making sure that you also focus on the topics you enjoy will help a lot with your workload. It's always a lot easier to work on topics you enjoy. It may be a welcome break to the difficult challenges of other topics, when you're still actually learning.

Ability to study something based on interest levels

Interesting   Uninteresting

27. Work on what you struggle with

It's also really important that you do work on what you struggle with. Making sure to understand your weaknesses and being able to focus your efforts on them is a skill itself. When you've found a topic that really doesn't click or doesn't make much sense at all, try to focus a bit of extra time on it compared to the others to build up your knowledge. This may be done by utilising the other tips in this book, such as finding videos on YouTube to give another perspective, speaking to other classmates and peers for some tips or even reaching out to the lecturer and asking for some 1-1 time to go through it to discuss and improve. There are always ways to learn, so don't feel disheartened if you really struggle with certain topics, you'll be able to make it through!

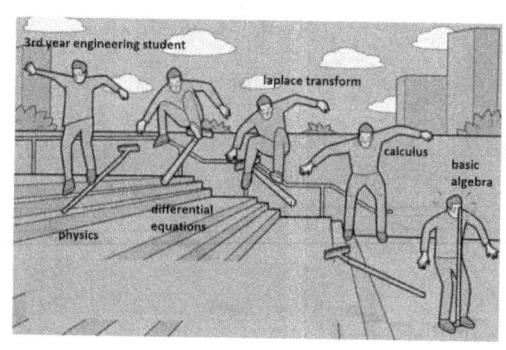

28. Don't always study alone

Now, this isn't always strictly true, sometimes it can be a godsend to study alone and get away from your peers. Maybe you're the type of person who studies better by themselves, I know I can be sometimes. However, studying as a group also comes with great benefits. If someone in your group is struggling you can help out, and teaching someone something really reinforces your learning for yourself which is always useful. At the same time, if you're struggling, someone else will help you out and help you to find the answer or explain a certain topic.

I had a group in university where we would sit on a group call and work individually, but if any of us started to struggle we'd just ask. It had the same effect of being in the same room but also gave a bit of flexibility.

29. Make study "cheat" sheets for your exams

Now I don't mean making cheat sheets to take into exams, that is cheating and I would never encourage or condone that! But, creating a 1 pager of information you may need to learn in an exam is really useful to have on hand. If you're completing example questions or past papers, its really useful to have the study sheet there. Try to get as far as you can without consulting the sheet and it helps test yourself with what you know, and you will begin to learn that whole sheet which will help you greatly during exams.

30. Everyone in your class feels the same as you

It's so important to remember that almost everyone feels the same as you in your class. Everyone is struggling and trying really hard to pass the course. Some people like to come across that they don't work hard or they barely study or revise, and that may have been possible during school, but during engineering it's simply not possible. Everybody is working hard and trying to pass their course. If it seems like some people barely do any work and are out all the time, that may even be reflected in their grades. Don't compare yourself to others as it's so hard to know what they're struggling with and working on.

31. Try to attend all lectures, but it's not the end of the world

Lectures are important, don't get me wrong, but they aren't the be all and end all of the engineering degree. Now some universities or colleges track attendance to lectures and can actually fail you if you don't show up, and in general I would agree that you should try and make it to every lecture you can. But, missing a few lectures here and there because you instead wanted to hang out with some friends or meet new people or maybe you're just massively hungover. Don't worry too much about it and just make sure that you catch up with the work in your own time instead.

32. Don't be competitive or toxic

Nobody likes that person after exams who immediately asks how the sam was and what specific answer you got for question 2.b. Chances are people just want to immediately forget it and move on until the results. Maybe you felt that the exam was super easy and you want to tell everyone about how easy it was, don't be that person! It's not going to help anyone and you'll just come across as a bit of a dick! Also, when exam results do come out, don't go around asking what everyone got. They may have struggled massively and don't want to share what they got and that's up to them. Don't feel the need to brag about your results.

33. It's normal to fall behind a bit

A lot goes on in engineering and there is a huge amount of work that needs to be done. Everyone falls behind a bit and struggles to catch up, I know I definitely did. If you do fall behind you might just need to put your head down and work an extra bit harder for the next few weeks to catch up. If you are really struggling if you fall behind, again, don't be afraid to reach out and ask for help.

Prioritise what needs to be done in what order based on deadlines as well. There's no point focusing on an optional tutorial if you have a massive deadline that needs to be completed in the next 2 days!

34. Get on LinkedIn and use it

LinkedIn can be a pain, but also massively useful. There's a lot of people on there who love to brag and show off about how great they are, and it does get quite annoying. But, it's great for networking and keeping in touch with a lot of people you meet and interact with. Get onto LinkedIn and set up your profile, initially it might be a bit empty with little experience, but that doesn't mean you can't connect with professors and people you might meet at networking events. These connections will stay on your LinkedIn and will always be there to reach out to and get in touch with in the future. You never know when they may come in useful! In the future, they may be working at the company you'd like to apply to and could give you a foot in the door.

Lumko Solwandle · 2nd
Transforming Infrastructure & Cloud at Nutanix
4d ·

No body:

LinkedIn Influencers:

Yesterday I was walking to an interview. There was a starving dog on the road. I stopped to feed him & missed the interview. The next day I got a call asking to come in to do the interview. I was surprised, but I went. Then the interviewer came in. He was the dog.

 3,408   113 comments

 Like   Comment   Share

35. There's a lot to be learned online

I know that I've mentioned YouTube already, but the wealth of knowledge to be found online is immense. If you're stuck on a certain subject or want to find out more, looking online for the answers will definitely bring you some useful information. It may be the exact topic taught in a different college or university which gives more information or the information in a different format.

Learning to google and search for the results as well is a highly useful skill.

36. It's sometimes possible to find PDFs of textbooks online

I'll leave this Wikipedia extract below for anyone that may be interested.

"Library Genesis (Libgen) is a file-sharing based shadow library website for scholarly journal articles, academic and general-interest books, images, comics, audiobooks, and magazines. The site enables free access to content that is otherwise paywalled or not digitized elsewhere. Libgen describes itself as a "links aggregator", providing a searchable database of items "collected from publicly available public Internet resources" as well as files uploaded "from users".

Libgen provides access to copyrighted works, such as PDFs of content from Elsevier's ScienceDirect web-portal. Publishers like Elsevier have accused Library Genesis of internet piracy. Others assert that academic publishers unfairly benefit from government-funded research, written by researchers, many of whom are employed by public universities, and that Libgen is helping to disseminate research that should be freely available in the first place" (Wikipedia, 2022)

Wikipedia. (2022). Library Genesis. [online] Available at: https://en.wikipedia.org/wiki/Library_Genesis [Accessed 10 Aug. 2022].

37. Create a portfolio of your work

This relates to the professional development tip that was discussed earlier. Creating a portfolio of the work you have completed will always come in useful when it comes to CVs, interviews and job applications, as well as when you apply for accreditation. Having a body of information that you can look at and use will always be useful.

It may seem difficult to stay on top of this portfolio, and is something that I still struggle with sometimes. It may be relevant to simply do it once every 6 months and go through your work, that way you can stay mostly on top of it without it taking up too much of your time.

38. Ask for help when you need it

This is an important tip and one that I can't emphasise enough. Please don't struggle alone. Always reach out to people who can help and speak to others who will be able to help you. This may be people in your university or college such as professors, or it may be administrative staff or maybe just some friends and family if you just want to have a chat. Speak to other people and don't struggle alone, it is a hard course but it's important to make sure you speak out when you need to.

Engineering Student Guidebook

39. WORD SEARCH BREAK

```
Z J Y E B I S Q U D C F W T X I L T E X P A D Y P R L J F I
V A D P O T D N I V F Y O I C J A R Q G S S B X G D E Q E J
V Q H H Q E X L O B M I H Q H J I N G O C R K H T Q T T J F
B D C F N D A P T I F G F O Y S R K C L E M C X L N O E N Z
O P A M U C O G J H B N T G O L E F D A B P Y Q W V S J G E
Y M Y N I N E C A P S O R E A W T X T A Z Z L V K N S I I N
M L M M F I E R W B P P N T V K A D H H A D O G V B M H S G
N E E B I H U F D M Y J Q E C N M U Z N D Y P N K V L G E I
C H O S X B J L J A F Y K C M Y M A Q N X B E F V D N M D N
C X N J P U B C W D J A S H G Q X G L Q T D R I M U K I A E
Y T A F S B I W Y O J K R N B Q W N E L D S A I E Q Z J J E
X F V L Q Y I Y J U T R G O Y W I W A B N F T F Z A E M G R
T J C I Q I G O H C I O N L X S O S P X D G I J F Z K A Q I
U E C L I P Q C M K K P I O C A T B L W S M O D I V Z F H N
Z D D C T O E N Y E F D N G O A P Y H H U J N S C L E F P G
L H Q R Y C S J A U D U I Y J F M V G E Z V Q Z V J J W C J
C A X B H A H D E W F I M N U X Z W L U J G F K Z S J C I M
O J R A C F G R M A C Q C J E Q O O C A T G L T I A S R S A
L C Q U O M A U X T I M A A V U R D W S J D Q P K I B Y T A
I G N U T B L E T J N V U G L T C T K N K R O J N B C Q I N
L J B G L C A F U R F H G U E Z E F J O A G K Q T L L P A O
F E S H V H U V P M S N H P P W T A K Z L W R T I A A U E I
J W J K G F Z R P O I B V K R J L R X Y O U W G X T C D E Q
Y H I J C U L R T K W T W Z E Z F C J M U W W P V Q I Z J U
L M I V B K U H R S U X D T K X L R A F A A P K T C N P S F
T I H O H U W O Q S K D Z E M U E I E F I C E J T O A J D G
B R S D K G W W I O S Z Q J E X H A H A O P F W C Z H X O B
I F M N O T C I V I L X U K Z B N B Z Y V K C H A Z C Y H X
L S G S E W I P L G P D A N Z P U W E Y O D B J M V E V N J
U R J N U X M O S S K U A S F N J B M C W C F P B O M I L H
```

ENGINEERING	AEROSPACE	OPERATION	CIVIL
STRUCTURAL	DESIGN	AIRCRAFT	MINING
NETWORKING	PETROLEUM	MECHANICAL	BIOMEDICAL
MATERIAL	CHEMICAL	TECHNOLOGY	

51

40. Learn and work on other soft skills

There's more to a well rounded engineer than just pure technical engineering skills. It's important to be able to work well in teams and communicate with other people. It's also important to show that you're a well rounded person that can work with people. I know some engineering students who got incredible grades and results, but were unable to succeed in interviews or jobs because they weren't well rounded and couldn't demonstrate any other soft skills.

There's lots of ways to work on these soft skills, such as joining societies, taking part in group projects and more. Generally, if you can notice on yourself that you have areas for improvements, that's always the biggest step towards working on those soft skills

41. There's more to life than being number 1

Being the best in your class can be a great feeling, but focusing all of your effort and work onto becoming the best and having that be your entire focus can be detrimental. If you took two graduates, one who was best in his class but had no other skills or experiences to his name, or the other who was maybe top 20 but was part of extracurricular clubs and took part in activities outside of just university work with some summer internship experience. It's likely that they'd look more favourably at the second more well rounded graduate, rather than the purely academic one. Make sure you enjoy your time!

42. Become S.M.A.R.T

S.M.A.R.T is an acronym used for goal setting for your career.

S - Specific
Objective clearly states, so anyone reading it can understand, what will be done and who will do it.

M - Measurable:
Objective includes how the action will be measured. Measuring your objectives helps you determine if you are making progress. It keeps you on track and on schedule.

A - Achievable:
Objective is realistic given the realities faced in the community. Setting reasonable objectives helps set the project up for success.

R - Realistic:
A relevant objective makes sense, that is, it fits the purpose of the grant, it fits the culture and structure of the community, and it addresses the vision of the project.

T - Timely:
Every objective has a specific timeline for completion.

43. Learn to code!

Coding will always be useful for anyone in engineering. There are thousands of resources available to look at and learn how to code and having a good understanding of the basics of coding will always be useful. In University it's likely there will be some coding courses available to learn how to code and I would definitely recommend making the most of them. I've listed a few of these resources below which may be useful for you.

Code Academy
Khan Academy
Code Avengers
Learn Python the hard way
Mozilla Developer Network
Code School
Treehouse
Udacity

44. Get hands on

Getting hands on experience is so important and actually making something with your own hands is so rewarding. It allows you to actually apply what you learn in the classroom into the real world. A few ideas of how to get some experience and get your hands dirty is below.

Solve real life problems: *There are ways you can help solve real life problems. Ideally, get in touch and work with professionals in your field of interest. They will be able to identify a real need in industry.*

Finish what you started: *Based on my experience, you will be hard pressed to produce a fully completed product within the time constraints of a 4-month course project. Sure, you might meet all your professor's expectations and get a good grade, but why stop there?*

Volunteer for extra projects: *Chances are there are teams in your university where you can volunteer to get extra experience, such as Formula Student. These are great ways to get experience with using your hands and actually building stuff.*

Make something fun: *Maybe you want to make a drone, or something fun you've seen on YouTube? Great idea! Have a go of making and you will learn an impressive amount!*

45. Attend those networking events and job fairs

We've all seen them and gone for the free goodies they give out! However, they're very useful for being able to chat to people who work in industry and get some insights into how they joined and any tips they can give. If you can connect with them on LinkedIn and drop them a message after the event and I'm sure they'd be willing to help and give you some advice on how you can improve your career.

Best Tips for a Career Fair

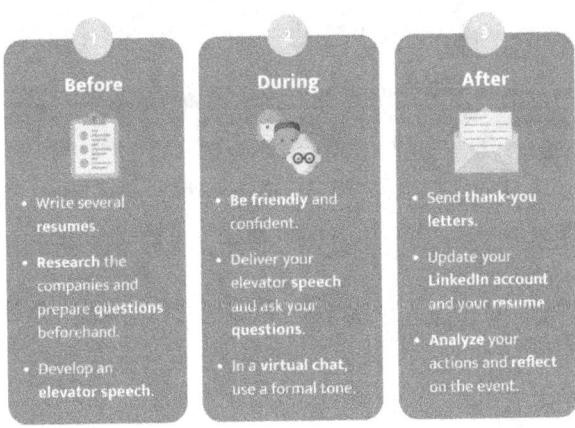

46. Apply for jobs early

Graduate programmes open up their applications really early, especially the highly regarded ones. Its definitely worth having your CV ready and up to date as early as possible, maybe even a year in advance of when you're looking to graduate.

If you are applying to a graduate programme, chances are they may have some sort of gamification tests to narrow down their application list. Try and get some experience of doing these tests and seeing what you need to improve on. Work with your university career teams as well to try and see if they can help.

It will be difficult finding a job but don't give up and make sure you get as much help and advice from everyone you can.

47. Practice good time management

Time management can be difficult, and its a good idea to get into good habits while you're at university in order to prepare yourself for the future. Time management itself is the process planning and controlling how much time to spend on specific activities. Good time management enables an individual to complete more in a shorter period of time, lowers stress, and leads to career success.

Here are a few tips to improve time management:

1. Set goals correctly
2. Prioritise effectively
3. Set a time limit for tasks
4. Take breaks between tasks
5. Organise yourself
6. Remove non-essential tasks
7. Plan ahead at the start of the day

48. You won't use everything you learn

You'll learn a lot during your studies, and there's a high likelihood that you'll only use 5% of what you actually learn in your engineering life! However, what University and College proves is that you are capable of learning these subjects to a high standard and makes you a well rounded individual who has a high level of understanding in lots of areas.

It will help you "think like an engineer" and begin to understand the engineering way of approaching problems. Nobody is expecting you to be an engineering genius when you graduate, don't worry!

49. Be proactive

Being a proactive student means doing things out of your comfort zone, going beyond the ordinary, and really putting yourself out there with every opportunity worthwhile and meaningful to yourself during your time at university. This is all simply through experience!

Every student should always think about how they can develop themselves or their knowledge of study every day. You can do this by getting real life experience and really getting yourself prepared for the real world. Gaining experience whether through a job, internship, volunteering, work experience is a vital and necessary skill and a great path towards opening many more opportunities into your lifetime.

50. Improve your communication skills

Engineers are sometimes notorious for their lack of communication. It's a very important skill and one that you should aim to improve whenever you can. Here are some tips on how you can improve your communication skills:

Make a good introduction: Before you can persuade anyone of anything, they need to know who you are and why they should listen to you. That will help build trust with your audience and give them a reason to take you seriously.

Balance detail with clarity: The only thing worse than listening to someone when you've no idea what they're talking about, is listening to them talk on about nothing at all. The key to avoiding either of these scenarios is working out what they you're speaking to already know and what they want to know.

Explain 'why' as well as 'what': When you're talking to non-engineers, it's important to understand that not everyone has the same enthusiasm for engineering that you do. What they're likely more interested in is what changes your engineering can make for them.

Be confident: You won't convince someone of your argument if you don't seem convinced yourself. Avoiding eye contact, mumbling and too much self-deprecation will make people think you don't know what you're talking about or don't believe it.

51. Work on your strengths

People always say work on your weaknesses, which is good advice, but I think it's equally important if not more important to work on your strengths. If you were an Olympic swimmer but your marathon running wasn't very good, when training for your next swim race you wouldn't prioritise running a marathon would you? No, you'd work on what you're good at and focusing on in order to improve and become better at your strength.

The same applies to engineering, if you're a really strong coder for example, focus on that as much as you can. Obviously don't neglect your other subjects to become well rounded, but don't feel bad for working harder on a certain subject!

52. Look out into the world of engineering

There is so much going on in the world of engineering that it's definitely important to try and keep up to date with all of the latest things. Every day there's some really interesting engineering projects being developed and it'd definitely a good idea to understand what's going on. That way you can build up a large base level of understanding of engineering going on in the real world that you can rely on and use in the future.

Here's a few websites to get started:

Interesting Engineering
The Engineer
IEEE Spectrum
Science Daily

53. Nobody is perfect, don't beat yourself up

It's great to have goals. Everyone needs to have things in life they strive to achieve. But is it necessary to constantly seek perfection in everything and then beat ourselves up for every misstep, or harshly self-criticise for each perceived underachievement? Perfectionism in its positive form can help us be more successful, but the negative or self-critical form actually impedes our progress.

1. Focus on positive self talk
2. Practice kindness towards yourself
3. Stop comparing yourself to others
4. Think of mistakes as learning opportunities
5. Be patient with yourself.

54. Don't be THAT engineering guy

We all know engineering is hard, but don't be that guy that goes around telling everyone how hard your degree it, or how easy there degree is in comparison! It won't win you any friends and will most certainly make you look like a bit of a dick.

Everyone's degrees/ courses and majors are difficult so and I know i definitely couldn't write a lot of the essays that are required in other courses!

When you're an engineering student at the end of a day and you forgot to tell someone how much harder your degree is than theirs

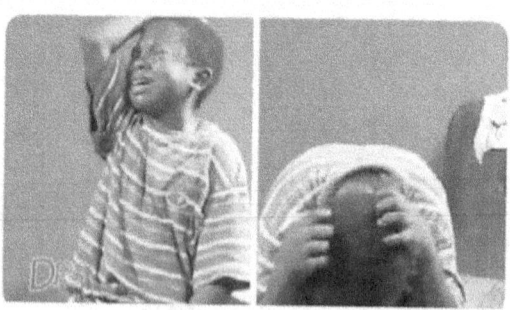

55. Try and create a study timetable during exam season

Exam season is difficult, and chances are you'll have a lot of exams to revise and study for. Everyone is different in the way they study, but I found the best way for me was to create a timetable of what I would study each day in the morning and afternoon. The hard part is then sticking to the timetable! I found this definitely helped me structure what to study and took the thinking out of each day of deciding what module to work on. I've copied below an example of what my timetables used to look like in the run up to exams.

Date of Monday	Week	Monday	Tuesday	Wednesday	Thursday	Friday	Saturday	Sunday
4/10/17	1	ACS	Thermo	Materials	Aero	Electrical	ACS	Thermo
		Aero	Electrical	Thermo	Thermo	Materials	Aero	Electrical
4/17/17	2	Materials	Aero	Electrical	ACS	Thermo	Materials	Aero
		ACS	Thermo	Materials	Aero	Electrical	ACS	Thermo
4/24/17	3	Electrical	ACS	Thermo	Materials	Aero	Electrical	ACS
		Materials	Aero	Electrical	ACS	Thermo	Materials	Aero
5/1/17	4	Thermo	Materials	Aero	Electrical	ACS	Thermo	Materials
		Electrical	ACS	Thermo	Materials	Aero	Electrical	ACS
5/8/17	5	Aero	Electrical	ACS	Thermo	Materials	Aero	Electrical
		Thermo	Materials	Aero	Electrical	ACS	Thermo	Materials
5/15/17	6	ACS	Thermo	Materials	Aero	Electrical	ACS	Thermo
		Aero	Electrical	ACS	Thermo	Materials	Aero	Electrical
5/22/17	7	Electrical Exam	ACS Exam	Aero	Aero Exam	Thermo	Materials	Thermo
				Aero		Materials	Thermo	Materials
5/29/17	8	Materials	Thermo	Materials	Thermo	Materials	Thermo	Materials
		Thermo	Materials	Thermo	Materials	Thermo	Materials	Thermo
6/5/17	9	Thermo	*Thermo*	Materials	**Materials Exam**			
		Thermo	Exam	Materials				

End of guide

Thank you for reading, I hope you found the guide useful for you and I'm sure that you will enjoy your time during your studies.

Author Bio

Thomas is currently a manufacturing engineer working in the Defence industry. He studied Aerospace Engineering at a British University that is in the Top 100 rankings in the world and is a Russel Group University. After graduating he has worked extensively as an engineer and wants to share his learning and experiences to help others.

www.ingramcontent.com/pod-product-compliance
Lightning Source LLC
Chambersburg PA
CBHW070310220526
45465CB00004B/1833